GOATS

JACK BYARD

KNOW YOUR GOATS

JACK BYARD

Old Pond
PUBLISHING

Know Your Goats

Old Pond Publishing is an imprint of Fox Chapel Publishers International Ltd.

Project Team
Vice President–Content: Christopher Reggio
Associate Publisher: Sarah Bloxham
Editor: Sue Viccars
Designer: Wendy Reynolds
Layout: John Hoch

ISBN 978-1-912158-48-5

A catalogue record for this book is available from the British Library.

Fox Chapel Publishing
903 Square Street
Mount Joy, PA 17552

Fox Chapel Publishers International Ltd.
7 Danefield Road, Selsey (Chichester)
West Sussex PO20 9DA, U.K.

www.oldpond.com

We are always looking for talented authors. To submit an idea, please send a brief inquiry to acquisitions@foxchapelpublishing.com.

Printed and bound in Malaysia
22 21 20 19   2 4 6 8 10 9 7 5 3 1

Cover photo: A thoughtful Angora goat.

Contents

Foreword

Goats are a diverse range of beautiful (and sometimes rare) animals. Bucks and does, billies or nannies, are the source of many a children's story (who hasn't read the Norwegian fairy tale *The Three Billy Goats Gruff* to their children?). Goats were taken into the human fold more than ten thousand years ago and most breeds have been wandering the fields, hillsides and mountains since time immemorial. From some we obtain fibre to create exquisite mohair and cashmere clothing. Others give us possibly more mundane (but equally important and extremely healthy) meat and milk, the latter frequently being turned into mouth-watering cheeses, bringing joy to me and a living to many farmers.

Goats are browsers and prefer unwanted brush, briar and weeds to grass, their lips and tongues selecting only the tastiest plants. Extremely intelligent and curious, they are also experts at escaping from the most secure fields.

In addition they make excellent pets: you may look twice when you see a goat wandering through a hospital ward, nursing home or rehabilitation centre, but therapy goats "bring love, affection, laughter and calmness to people ailing in mind, body or spirit".

Jack Byard, Bradford, 2019

NOTE: All weights, sizes and measurements in this book are averages based on breed records and the terrain in which the animals live. Male goats are known as billy or buck; female goats as nanny or doe.

(opposite) Angora goats.

1

Anatolian Black

Characteristics

Weight: 99–198 lb
(40–90 kg).
Height: 28–40 in
(70–100 cm).
Both billies and
nannies have
curved horns.

The Anatolian Black has been domesticated and bred on small farms in what is now Turkey since 700 BC. It is described as the Syrian type, with long droopy ears and long hair.

This calm gentle mountain breed is normally found in large herds around the Mediterranean and Aegean regions and is well adapted to survive the wild weather and sparse feed; the long, thick, hairy undercoat insulates the animal against the cold. They are occasionally brown, grey or pied, and have a tremendous tolerance to disease. They breed all the year round, feeding on grass and small pine, olive and almond saplings and grain. If there is plenty of food available for the goats the breeder will set up camp along with the herd.

The Anatolian Black is also a brilliant mowing machine. Some say the breed is aggressive and dangerous, but experts confirm that this is not the case. The Anatolian Black is bred for its milk and meat and for its fibre: it sheds 1 lb (0.45 kg) a year.

2

Angora

Characteristics
Weight: male 180–225 lb (82–102 kg), female 100–110 lb (45–50 kg).
Height: male 48 in (122 cm), female 36 in (92 cm).
Both have gently curving horns.

Goat hair or fibre has been used to make clothing for over three thousand years. Mohair is the silky, lustrous and hard-wearing fibre from the Angora goat, which is shorn twice a year, producing 9–11 lb (4–5 kg). Originally coming from the Himalayas, the goats were herded to Ankara by Suleiman Shah while fleeing from the legendary Genghis Khan. "Angora" derives from Ankara, and "mohair" from the Arabic *mukhayua*.

The finest mohair comes from the six-month-old kid, coarsening as the animal gets older. The main colour is white, but black to grey and silver are being bred, plus reds and brown. Mohair is an all-season fibre, cosy in winter and cool in summer. Hats, scarves, socks, fleeces and suits – and I must not forget the cuddly teddy bear. Records show that the Angora, with its long droopy ears, first reached British shores in the 1500s, but did not survive, and that the Angora owned by Queen Victoria suffered the same fate.

They set foot in Australia in the 1830s, and arrived in the United States in 1849. It was not until the 1980s that they were truly established in the British Isles.

3

Characteristics

Weight: male 143 lb (65 kg), female 99 lb (45 kg). Height: male 29½–33½ in (75–85 cm), female 27½–31½ in (70–80 cm). Usually without horns.

Appenzell

This is a beautiful pure white dairy goat: a native of Switzerland usually found in the cantons of Appenzell and St Gallen, and a rare and endangered breed dating back many hundreds of years.

In the 1980s it was infected by a terrible disease, Caprine Arthritis Encephalitis (CAE), which forced it to the point of extinction. In 2000 only 677 were registered, and in 2011 this had increased to 1,479. This was due to the help and support of the FOA, the Swiss Goat Breeders Association and ProSpecieRara, the Swiss Preservation Foundation which supports the breeding and cultivation of traditional animals and crops.

They are bred mainly for their milk, which contains 2.9% fat and 2.7% protein, a perfect base for cheese-making. In close cooperation with ProSpecieRara and the Coop supermarket of Switzerland, innovative cheese-maker Matthias Koch has created a new cheese made solely from Appenzell goats' milk, which has been extremely successful. With a guaranteed market for the milk the farmers are enlarging their herds to keep up with increasing demand. The beautiful Appenzell is now out of danger.

4

Arapawa

Characteristics
Weight: male 130 lb
(59 kg), female
59–79 lb (27–36 kg)
Height: male 26–30 in
(66–76 cm), female
24-28 in (61–71 cm).
Billies have flattened,
sweeping horns; the
nannies' horns are
shorter, rounder and
curve backwards.

The Arapawa arrived in New Zealand courtesy of the British explorer Captain James Cook, landing on the shores on 2 February, 1773; in 1777 he presented a Maori chief with a further two goats. It is accepted that the Arapawa are the descendants of the Old English Milch goats. In 1970 the New Zealand Forest Service decided to cull them, bringing them to the brink of extinction, since they "believed" they were damaging ancient woodland. Local farmer Betty Rowe spent a lifetime battling on their behalf. A number were taken to safety off the island to breed elsewhere.

In 1993 goats from the Betty Rowe Sanctuary arrived in the USA, and in 2004 six arrived in the UK. There are less than five hundred domesticated Arapawa worldwide. The American Livestock Conservancy believe them to be one of the rarest breeds in the world. They have short fluffy coats with shaggy leggings. The colours vary from black, brown, tan and fawn to creamy white and tri-coloured, with black or dark brown badger stripes on the face.

5

Bagot

Characteristics
Weight: male 103 lb
(47 kg), female
80 lb (36 kg).
Height: male 26 in
(66 cm), female
23 in (58 cm).
Both have long horns,
twisting and sweeping
backwards.

The Bagot, possibly Britain's oldest breed of goat, is strongly believed to have been brought to these shores by the Crusaders. They were presented to Sir John Bagot in the 1380s by King Richard II in gratitude for the hospitality he had received at Blithfield Hall in Staffordshire. These semi-feral goats have browsed the parklands there for over six hundred years. In World War II the herd was found guilty of damaging vital crops and sentenced, by the War Agricultural Executive, to be destroyed. It was eventually agreed to reduce the herd to sixty, and that number was retained for the remainder of the war.

A number of black-and-white goats wander the hills of Wales; these are not Bagots, but escapees from the Hall. Commercially the Bagot has nothing to offer. The head and shoulders are black and the rest of the body is white. They are excellent conservation grazers, but are bred for their beauty and rarity.

6 Bilberry

Characteristics

Weight: 132–165 lb
(60–75 kg). Height
information was
unavailable at
time of writing.
Both billies and
nannies have very
large curling
wavy horns.

This is a very rare, unique, ancient and gentle feral breed, very different from any other breed in Ireland. DNA testing suggests that it could descend from the Asian Pashima Down breed. The Bilberry is large with a long, shaggy, silky coat, a long beard and a fringe that covers the eyes.

They are thought to have been brought to Waterford Quays in Ireland by the Huguenots escaping religious persecution in Europe; in the seventeenth century they were put out to graze on common land at Bilberry Rocks, but have been moved along many times for many reasons (at one time people simply did not want the goats living there). By 2000 only seven remained, but with the help of Martin Doyle, the organizer of the Bilberry Goat Heritage Trust and the Irish Wildlife Trust, they now have a permanent home where they can live undisturbed, and numbers have risen to 89. This land has recently been bought by a property developer, but the minerals in the rocks are essential for the breed. Economic considerations must not be allowed to adversely affect the survival of an important piece of Irish Heritage.

7

Bionda dell' Adamello

Characteristics

Weight: male 154–165 lb (70–75 kg), female 121–132 lb (55–60 kg).
Height: 29½ in (75 cm).

The Bionda dell'Adamello ("blonde mountain goat") originated in the Valle Savoir region of Lombardy in Italy, although its ancient origins – before it spread out into the neigbouring valleys – are a little vague. Paintings of the breed date back to the eighteenth century, and in the mid-twentieth century the Bionda was on the verge of extinction. Numerically it is the most successful breed: in 1995 there were barely one hundred, but with the help of the farmers and the R.A.R.E. Association the numbers have risen to over four thousand and are continuing to improve.

It would be tempting to intensively breed the Bionda to take advantage of the increasing popularity of the smoked Fatuli and Mascarpi whey cheese made from their milk, but the Alpine breeds do not lend themselves to intensive breeding. In spring and summer they are taken up to the mountains and allowed to graze, as nature intended. They are capable of surviving in most climates. The long, fine fibre is light brown with regular patches of white, and white stripes each side of the muzzle.

8

Boer

Characteristics

Weight: male 240–300 lb (110–135 kg), female 200–220 lb (90–100 kg). Height: 29½–31½ in (75–80 cm).

The Boer (the Dutch word for farmer) was originally developed in South Africa in the early twentieth century. The Boer is believed to have been created using goats from the Namaqua Bushmen and the Fooku tribe and crossing and improving them with European and Indian breeds. It is the only goat breed in the world to be bred specifically for meat and in appearance is entirely different from the dairy breeds. The Boer has a short stocky body with a broad chest, and usually has short, white, smooth hair with a chestnut-brown head and floppy ears.

The Boer was imported into the British Isles in the mid- to late-1980s, and despite being bred in a warmer climate has adapted well to the vagaries of the British weather. It is now well established, and has created an excellent export market. It will quickly clear a pasture of weeds, thus improving it for grass-feeding stock.

9

British Alpine

Characteristics

Weight: male 170 lb
(77 kg), female
135 lb (61 kg).
Height: male 37 in
(95 cm), female
32 in (83 cm).
Both are horned
or polled.

There are two types of Alpine goats: the British and the French. A Swiss goat with the grand name of Sedgmore Faith at the Paris Zoo had the beautiful and distinctive black-and-white markings of her breed; in 1903 she was brought to England and crossed with a Toggenburg of similar colouring. All the kids had beautiful Swiss markings. More breeding and refining took place; the British Alpine had arrived and was here to stay, Sedgmore Faith being the grandmother of them all.

The British Alpine is tall and graceful with a short, fine, glossy black coat with white or cream facial and leg markings. The ears are erect, pointing slightly forward. An area to forage, a supply of hay and a muesli-type supplement is the key to happiness. They produce a good supply of quality, easily digestible alternative milk for all the family.

10

Characteristics

Weight: 100–200 lb
(45–54 kg).
Height: male 24–
27 in (61–69 cm),
female 22–24 in
(55–61 cm).
The large
horns form a
scimitar twist.

British Primitive

The British Primitive is the name that covers the breeds previously known as the Old English, the Scottish, the Welsh, the Irish and British Landrace, not forgetting the Old British Goat. The British Primitive is a descendant of the goats bred by the farmers of the Neolithic (New Stone Age). The Vikings, Saxons or Celts would have bred this hardy animal that was a born survivor, protecting itself and its young against predators as well as having the ability to survive the harshest of weather on a poor and meagre diet. They have long, thick, dense hair, coloured mainly white, grey and black.

The breed provided the farmer and his family with milk, meat, skin and fibre for clothing, and tallow for heating and lighting. Nothing was wasted. They are now used for scrub clearance and conservation grazing.

11

Characteristics

Weight: male 140–190 lb (64–86 kg), female 120–140 lb (55–64 kg). Height: 22½–31½ in (65–80 cm). Both are horned or polled.

Brown Shorthair

The Brown Shorthair is a popular breed in the Czech Republic and Germany. It was developed in the late nineteenth and early twentieth centuries on the borders of the two countries using local brown goats and Brown Alpines. Great importance was placed on maintaining its hardy and adaptable traits with the ability to survive in the worst of climatic conditions. It has a short, glossy brown coat with a black spinal stripe running from a triangle behind the ears to the base of the tail. The underbody, lower legs, hooves and inner side of the upright ears are black. In winter they can shelter in barns when temperatures can drop to -20°C (4°F). In summer they typically browse at 2,600 feet (800 m), feeding on bushes and grass; a great favourite is raspberry canes.

The Brown Shorthair is a high-yield commercial dairy goat; the milk has many uses, including cheese-making.

12

Chamois

Characteristics

Weight: male 66–132 lb
(30–60 kg), female
50–99 lb (25–45 kg).
Height: 28–30 in
(71–76 cm).
Both have horns.

The beautiful Chamois originated in the Canton of Berne in Switzerland and it is at home in the steep and rugged terrain of the Alps. This nimble, surefooted animal can reach speeds of up to 10 mph (16 kmph) in these rocky landscapes. The summer coat is reddish brown with a dark dorsal stripe; in winter the coat is blackish brown. They have a brown face with a darker stripe running from the muzzle to the eyes.

The Chamois is an ideal dairy goat and produces good quantities of sweet-tasting milk; their docile temperament makes them ideal pets (until you try to trim their hooves, when they can be quite a handful!). They are said to be brilliant bramble mowers. Milk from the Chamois is used to create the beautiful and creamy Fryberg-Chäs cheese.

13

Dutch Landrace

Characteristics

Weight: male 80 lb
(36 kg), female
60 lb (27 kg).
Height: male 25 in
(64 cm), female
24 in (61 cm).
Both have horns
which arc backwards
with an outward twist.

The Dutch Landrace is similar in many ways to the other breeds of North West Europe. The breed has been used to develop and improve many others. Over the years the bloodline became weaker; by 1958 only two animals remained and it was left to the enthusiasts to come to the rescue. The breed is famous for its long, multi-coloured hair, and the most common colours are brown-black, blue-grey and white. Capable of surviving and thriving in most climates they are, without doubt, one of the most beautiful goat breeds.

The milk yields are not sufficiently high for the Dutch Landrace to be farmed commercially, but the quality milk is used for cheese-making. Herds of Dutch Landrace are used on National Nature Reserve land to keep the moors and open grassland areas free of trees.

14 Fainting

Weight: 80–150 lb
(36–68 kg).
Height: 17–25 in
(43–64 cm).
**Both billies and
nannies usually have
horns which sweep
upwards and outwards.**

In the early nineteenth century an elderly farm labourer, strangely dressed, arrived in Marshall County, Tennessee accompanied by four goats and a cow: enter John Tinsley. Soon everyone was talking about John's goats which, if surprised or frightened, would stiffen or fall over. The condition is known as myotonia congenita: the muscles contract for a few seconds (this neither harms nor hurts the animal, which will continue to chew food already in its mouth). Before moving on John sold his goats to Dr H. H. Mayberry, who bred them and sold them locally. They became known as the Tennessee Fainting goats.

The Fainting goat's coat comes in many colours but mainly black and white, or red and white, and can be short to shaggy. They take up to three years to mature and make ideal pets.

15

Characteristics

Weight: male 143 lb (65 kg), female 101 lb (46 kg).
Height: male 33½ in (85 cm), female 31½ in (80 cm).
The upright corkscrew horns are up to 27½ in (70 cm) long.

Girgentana

The ancestors of this ancient breed are believed to have originated from Afghanistan and Baluchistan; they arrived in Mazaro, Sicily when the Arabs invaded in AD 827. The breed prospered and slowly spread throughout the island. It is mainly found in the province of Arigento, Girgenti in local dialect, from where the breed derived its name. The Girgentana is all white, with occasional brown spots on the face.

Because it is less productive than modern breeds it is at serious risk of extinction. In the mid-twentieth century there were in excess of thirty thousand animals, but in 2013 only 390 heads were registered. Work is underway to ensure the survival of the breed. Drinking the milk, or buying the quality cheeses made from it, help support the farmers who are their guardians, and is a great way to save them. Traditionally the milk was given to children and the elderly.

16

Characteristics

Weight: male 150 lb
(68 kg), female
120 lb (54 kg).
Height: male
28 in (71 cm),
female 30 in (66 cm).
Billies can be horned.

Golden Guernsey

The origin of the breed is unknown, but research by the University of Cordoba in Spain has concluded that the breed is indigenous to the island. Even during the dark days of World War II, when for five years Guernsey was under German occupation, the breed was still being registered. The long-haired coat comes in all shades of gold from pale to bronze, sometimes with small white markings and a star on the forehead.

One of the largest and possibly best-known herds was that of Miss Milbourne of L'Ancresse who owned over fifty Golden Guernseys, which played an important part in helping revive the breed in the 1930s. The gene pool is very small and tremendous efforts are being made to improve the situation and secure the future of this beautiful breed. In an effort to safeguard their future a number are being raised in New York State, but only the purebred with UK registration are allowed to use the proud title of "Golden Guernsey".

17

Icelandic

Characteristics

Weight: male 200–220 lb
(60–80 kg), female
150–161 lb (35–60 kg).
Height: male 29½–31½ in
(75–80 cm), female
25½–27½ in (65–70 cm).
Both can be horned
or polled.

The Icelandic goat is an endangered species, and there are no pure Icelandic goats outside the country. It is also called the Settlement goat, arriving in Iceland with the Norwegian settlers over eleven hundred years ago, and there have been no further imports since that time. They are brown, grey, black and white in various patterns. Under the coarse outer coat of long guard hair is a coat of high-quality cashmere that is combed out once a year, harvesting 6–8 oz (170–225 g). In 1986 six were exported to Scotland for a cashmere breeding programme. They were put together with three other breeds to produce a new synthetic goat breed, "The Scottish Cashmere Goat".

In the early twentieth century there were in the region of three thousand goats, but by the latter end of the century the number had plummeted to fewer than four hundred. The government has introduced a conservation programme, and the numbers are slowly increasing. There is commercial and economic potential for the Icelandic through cashmere, milk and meat. They are at present kept as pets.

18

Kiko

Characteristics

Weight: male 250–300 lb (113–136 kg), female 100–150 lb (45–68 kg). Height: 17–25 in (43–64 cm). Both have horns; the billy's impressive spiralling horns sweep upwards and outwards.

The Europeans who discovered New Zealand in 1769 brought goats with them, and a number of them escaped and thrived. These goats were to become the modern Kiko, growing in number through having no natural predators. With no shelter, no supplementary feed, no veterinary care and no help giving birth, they became a tough and hardy breed, resistant to disease, parasites and the weather. The Kiko, Maori for meat, became totally self-sufficient.

The breed as it is known today was developed by Garrick and Anne Batten on the South Island in the 1970s, and only the best of the best were selected. In the early 1990s a number were imported into the USA by Dr An Peischel where they continued to grow in popularity. In summer the coat is smooth and shiny and usually white, although it can be coloured; in winter the hair is long and flowing.

19

Kinder

Characteristics

Weight: male 135–150 lb (61–68 kg), female 110–125 lb (50–75 kg). Height: male 20–28 in (51–71 cm), female 20–26 in (51–66 cm). Both are horned.

The Kinder (KIN-der) is a cross between a Nubian and a Pygmy goat. On the Zederkamm Farm in Washington in 1985 the Nubian buck died leaving the owners, Pat and Art Showalter, with two Nubian females but no mate. They did not want to send the goats to another farm to mate. Pat and Art also bred Pygmy goats… Left to his own devices the Pygmy buck accomplished two successful matings, making use of logs and rocks and the sloping land to gain the correct height. In late June 1986 three Kinder does were born, and a year later the first Kinder buck.

The coat is short, fine hair in any goat colour and pattern; the longish ears stick out to the sides. The Kinder is gentle and family-friendly and makes a good pet. The milk is ideal for drinking and for making delicious yoghurt and cheese.

20

La Mancha

Characteristics
Weight: male 150 lb
(70 kg), female
130 lb (59 kg).
Height: male 30 in
(76.25 cm), female
28 in (71 cm).
Both can be horned
or polled.

The first La Mancha (or Lamancha) were bred in Oregon by Mrs Eula Fay Frey in 1938. The two offspring, Peggy and Nesta, are the foundation of the breed. It is also the only goat breed developed in the USA. The breed is recognizable by its very short ears, less than 1 inch (2.5 centimetres) long. Short-eared goats, found throughout Spain, are mentioned in ancient Persian writings, and also by the Spanish missionaries who colonized California in the mid-eighteenth to mid-nineteenth centuries, bringing with them short-eared goats – and it is believed these are the ancestors of the La Mancha.

They are known for their high production of butterfat-rich milk which is used for making cheese, yoghurt, ice cream and soap. They have a laid-back attitude to life, are hardy and make superb brush and bramble mowers. The hair is short, fine and glossy. And what about colour? Think of any goat colour or combination of colours – and that is the La Mancha.

21

Characteristics

Weight: male 159 lb
(72 kg), female
79 lb (36 kg).
Height: male
28 in (72 cm), female
26 in (67 cm).
Both are horned.

Messinese

The Messinese is an indigenous goat found in the mountainous regions of Peloritani and Nebrodis in the provinces of Messina and Catania in Sicily. They are well adapted to living in this hilly and mountainous region where the tendency is to let them roam free in an almost semi-feral state. The Messinese feeds on the sparse mountain vegetation, browsing on basil, fennel, sorrel and rosemary which flavour an excellent range of cheeses produced from the milk.

It has long hair; it can be pied or streaked white, brown, black or red in varying shades. It is bred especially for its milk, in an area that has an ancient tradition for making cheeses. Artisan cheese-makers pass the skills and techniques from father to son.

22

Nigora

Characteristics

Weight: male 80–120 lb (36–55 kg), female 60–90 lb (27–41 kg).
Height: 19–29 in (48–74 cm).
Both can be horned or polled.

The Nigora is a small- to medium-size goat developed in the USA in the 1990s as a dual-purpose animal to produce quality fibre and milk. The breed is a cross between a Nigerian Dwarf buck and an Angora doe (the first doe was named Cocoa Puff of Skyview). Since then there have been gentle adjustments and improvements to the breed. It was designed with the urban hobby farmer and smallholder in mind, its small size and relative ease of care also making it an ideal pet. The advertising blurb states "It is a breed particularly suited for the micro-eco farming niche." The breed can be any colour and markings, and has erect ears.

The fibre is classed mainly as "cashgora", a cross between cashmere and Angora. The fibre is in three types: (A) the Angora type, long and lustrous; (B) the cashgora combines mohair with the cashmere-type undercoat and is of medium length; (C) like cashmere and is shorter. What more could you ask for: quality milk *and* the makings of a sweater!

23 Nubian

Characteristics

Weight: male 175 lb
(79 kg), female
135 lb (61 kg).
Height: male 35 in
(89 cm), female
30 in (76 cm).
Hornless by disbudding.

The Nubian goat is also called the Anglo Nubian, but in the USA is just known as the Nubian. The breed was developed in the UK in the nineteenth century from breeds originating in the Middle East and North Africa and the Old English Milch goat; small improvements have been carried out over the years, the Anglo-Nubian officially being recognized as a breed in 1896. It has a short, fine glossy coat which can be of any colour or pattern, large pendulous ears and a Roman nose.

The Nubian is also known for producing high-quality, high-butterfat milk, the "Jersey" of the dairy goat world. Being an extremely calm and affectionate breed, a pat on the head or stroking the neck will make you a new friend – they thrive in human companionship. They will bleat to call you, especially when they want food. They require plenty of roughage, branches, weeds, the occasional rose clippings and shrubs. If you are considering keeping a Nubian, remember that "Houdini" is their middle name…

24

Oberhasli

Characteristics
Weight: male 143–165 lb (65–75 kg), female 99–120 lb (45–54 kg).
Height: male 30–34 in (75–85 kg), female 28–32 in (70–80 cm).
Both are polled.

The Oberhasli goat is a dairy breed originating in the mountainous canton of Bern in Switzerland. Oberhasli goats were first imported into North America in the early 1900s, though it was not until 1936 that purebred herds were established and maintained. The Oberhasli is alert in appearance with a friendly, quiet and gentle disposition. While the does are a dependable source of milk, bucks and wethers are also useful as pack animals because of their strength and calm demeanour.

Oberhasli goats are brown, with hues between light tan and deep reddish brown. Two black stripes from the eyes to the black muzzle give a distinctive facial appearance. The Oberhasli has a black belly and a light grey to black udder. The legs are black below the knees, prompting the Swiss to refer to them as "booted goats". The breed is well known internationally, and is relatively numerous in Switzerland.

25

Peacock

Characteristics
Weight: male 165 lb
(75 kg), female
123 lb (56 kg).
Height: 28½–31½ in
(73–80 cm).
Both have large
horns, curving
backwards.

The Peacock goat is found in the cantons of Graubunden and upper Tessin in Switzerland, and was "discovered" in 1887. The name came about because of a journalist's slip of the pen: "he" should have written *Pfavenziege* (striped goat), but Peacock is more suitable for this beautiful animal.

Its true origins are unknown, but it has been around in the Swiss, German and Austrian Alps for many years. The thick coat is mainly white, with a black rear end and boots. The face has black stripes and spots. Blood tests carried out in the 1930s put the Peacock as a descendant of the Grisons Striped goat, but more recent tests could not confirm this. The Peacock is not officially recognized as a breed and therefore does not receive any public funding, tempting breeders to switch to other breeds. Why should the stroke of a pen decide an animal's right to life? Docile, agile and hardy, well able to survive the mountain weather and meagre pasture, many associations are actively promoting the breed to ensure it does not disappear into the Alpine mists.

26

Poitou

Characteristics
Weight: male 120–165 lb
(55–75 kg), female
105–140 lb (40–65 kg).
Height: male 30–35 in
(75–90 cm), female
25–30 in (65–75 cm).
Both can have
curving horns.

The legend has it that the Poitou dairy goat was left on French soil by Arab warriors after their defeat in AD 732, but archaeological evidence suggests that they have been in the region for over five thousand years. The goat we know today originated around the area of the Sevre River in the 1800s. In 1876 the crops failed and the farmers turned to dairy production to make a living. In 1906 a cooperative was formed and local cheeses entered the market. In 1920 a Foot and Mouth outbreak reduced the numbers. In 1985 a major local agricultural college decided to replace its herd of Poitou, creating an outcry from the local breeders who set up a protection society – and the numbers have been on the increase ever since.

Poitou goats have a distinctive appearance. They are tall, with long, shaggy hair, and black-brown with white marks on the head and neck. They are peaceful, calm and friendly. The milk is used to produce the famous Chabichou du Poitou cheese, among others.

27

Characteristics

Weight: male 60–86 lb
(27–39 kg), female
53–75 lb (24–34 kg).
Height: 16–23 in
(41–58 cm).
Both are horned.

Pygmy

The Cameroon Dwarf goat, as it was originally known, comes from West Africa. During the 1950s many were imported into mainland Europe where they were exhibited in zoos as exotic animals; within a few years they had found their way to the British Isles where they have proved very popular, and were then exported to the USA in the 1950s.

The Pygmy is hardy and adapts to most climates. They are great browsers and will clear your weeds (and no doubt your herbaceous borders). Small they may be, but they produce a large amount of milk. These goats are mainly black, white, grey and brown in colour and both sexes have horns, but only the billy has a beard. They are social animals and need a companion, not necessarily one with four legs: the Pygmy is an ideal pet for children of all ages. They enjoy climbing and something to jump on and a warm, draught-free home in the winter. Because they are gentle and affectionate they are frequently used as therapy animals. They are not bred for meat or milk, but purely for pets and companionship.

28

Pygora

Characteristics
Weight: male 75–95 lb
(34–43 kg), female
64–75 lb (29–34 kg).
Height: male 23 in
(60 cm), female
18 in (45 cm).
Can be horned or
polled.

The Pygora originated in the USA. The inspiration for this beautiful breed came from Katherine Jorgensen seeing the coloured curly goats on a visit to the Navajo Indian Reservation. The first generation, a cross of the Pygmy and Angora, created in the mid- to late-twentieth century, were all white; the colours – red, brown, black and grey or a mixture – did not appear until the second and third generation.

Many artists use this exquisite fibre for hand- and machine-spinning, knitting, weaving and creating tapestries. There are three types of fibre: (A) is an angora style, being lustrous, curly and up to 6 inches (15 centimetres) long; (B) is curly and can be lustrous or matte and 3–6 inches (7.5–15 centimetres) long; (C) is a cashmere style, 1–3 inches (2.5–7.5 centimetres) long, almost straight, with a matte finish. A and B are shorn twice a year; C is harvested by brushing. Most Pygora, because of their docile and friendly nature, are kept as pets.

29

Rove

Characteristics

Weight: male 80–90 kg (176–220 lb), female 110–132 lb (50–60 kg). Height: male 32 in (81 cm), female 29 in (74 cm). Both have long twisting horns.

It is believed that the Rove arrived in Marseille in 600 BC courtesy of the Phoenicians after one of their ships foundered and a number of goats swam ashore. Here they were developed and improved by the local farmers, and eventually named Rove after the village on the outskirts of Marseille. The breed's most striking feature is its long twisting horns that can grow to 4 feet (1.2 metres) in a mature adult.

The Rove wanders the Alpine countryside that abounds with aromatic herbs such as citronella, thyme and rosemary. These delicate flavours are to be found in the local goat cheeses. In the 1970s the breed was on the verge of extinction, but is now under the protection of the Association de Defense des Caprine du Rove and is well on the road to recovery with numbers exceeding six thousand.

30

Characteristics

Weight: male 143–176 lb (65–80 kg), female 121–143 lb (55–65 kg). Height: male 37 in (94 cm), female 32 in (81 cm). Both usually have horns.

Saanen

The Saanen's origins are in the Saanen valley in Switzerland. In the late nineteenth century many thousands were rounded up for export. The breed arrived in the USA in 1904; they were also distributed throughout Europe, arriving in the British Isles via Holland in 1922.

The Saanen is the largest of the dairy goats and can produce up to 1 US gallon (4 litres) of milk a day, gaining her the title of "The Queen of Dairy Goats". Like all goats they need a reasonable space to browse eating leaves and clover. They do not like getting wet, so a shelter from the rain is a must, and a good warm draught-free shelter for the winter. The hair is short and white or light cream, some having a fringe down the spine and longer hair over the tops of the legs, and the ears are erect and pointing forward. This calm, gentle, easy-to-handle animal makes an ideal pet and companion for children.

31

San Clemente Island

Characteristics

Weight: male and female
51–121 lb (23–25 kg).
Height: male 23½ in
(60 cm), female
22½ in (56.75 cm).
Both are horned, their
large horns resembling
those of the Spanish
goat and flaring up
and outwards.

The goats were brought to the island, which lies off California, by Salvador Ramirez in 1875. The US Navy is responsible for the upkeep of the island. In 1980 there were in excess of fifteen thousand goats browsing there, damaging plant and animal life. The courts sent in trappers and three thousand were re-homed off the island. Further controls were needed, and so the Navy decided to go in with helicopters and guns, but this action was blocked by the Fund for Animals.

There are now only seven hundred San Clemente goats. A study in 2007 found that the goat was not of Spanish origin, but a genetically distinct breed and unrelated to the numerous other breeds in the study. Colouring is light brown to dark red or amber. The head is black, with a brown stripe from around the eye to the muzzle, and a dorsal stripe down the back. These are shy gentle creatures that make ideal pets and produce sweet-tasting milk.

32

Savanna

Characteristics

Weight: 200–250 lb
(91–113 kg).
Height: 19–25 in
(48–62 cm).
Billies and nannies have
black horns growing
back and downwards.

The Savanna was developed in Douglas, South Africa, in 1955 by the Cillier family using local goats. To survive in the harsh local conditions the goats needed to be tough, strong, disease- and parasite-resistant. They needed to be able to survive extremes of heat and cold, intense sunshine and rain. In the cold, fine hair grows to give added protection. The coat is short, smooth, white hair, occasionally with red, blue or black "freckles". The skin is loose and black, and the ears floppy.

Their diet is not that of the usual farm animal: they eat large bushes, trees and seedpods. The main requirements to survive on this diet are strong jaws, long-lasting teeth and strong back legs to enable them to reach the higher leaves and branches. If you want a breed of goat that doesn't require a wet nurse the Savanna is for you.

33

Characteristics

Weight: male 150–200 lb
(68–91 kg), female
80–135 lb (36–61 kg).
Height: 17–25 in
(43–63.5 cm).
Both have horns; those
of the billy may be
large and twisted.

Spanish

In the sixteenth century Spanish explorers brought goats from their homeland to the Caribbean Islands, and eventually they arrived in what is now the USA and Mexico. The Spanish goat of today is the result of cross-breeding with many of the New World breeds. Until *The Mayflower* sailed from Plymouth and anchored off what became New England in 1620 the only goats in North America were the Spanish.

It is an extremely hardy breed, but unfortunately it is threatened with extinction and is on the American Livestock Breeds Conservancy watch list. Like most goats they were originally bred for food but their other names – the Brush or Scrub goat – give a clue: someone once wrote, "They are excellent for clearing brush and undesirable plant species from pastureland." All colours are acceptable. The hair is short with longer hair on the lower parts of the body. The ears are long and fall to the sides of the head.

34

Characteristics

Weight: 110–176 lb (50–80 kg). Height: 26½–33½ in (67–85 cm). The horns are thick and curve back from the head; the nanny's (though similar) are slightly smaller.

Stiefelgeiss

The Stiefelgeiss, the Booted goat, is a robust and hardy breed and well suited to life in its harsh mountain habitat. It could until 1920 be found in the uplands of St Gallen in Glarus, Switzerland, but by the 1980s the breed was on the verge of extinction, and so ProSpecieRara took control. The Booted Goat Breeders Club of Switzerland has now taken over management of the breed and farmers all over Switzerland are being actively encouraged to breed the Stiefelgeiss for its milk, meat and fibre.

Their appetite for leaves, buds and bark makes them an ideal tool to preserve the quality of pastureland. They also make ideal surrogate mothers. Their colour ranges from a light greyish brown to a dark reddish brown. They have beards and longer hair on their back legs which is usually a different colour, hence the "boots".

35

Toggenburg

Characteristics

Weight: male 150–200 lb (68–91 kg), female 125 lb plus (57 kg plus). Height: male 34–38 in (86–96.5 cm), female 30–32 in (76–81.25 cm). Usually without horns.

They were developed three hundred years ago in the Toggenburg region of St Gallen in Switzerland and were the main source of income for the poorest families. All the Toggenburgs, called Toggs by their loving owners, were pooled and grazed on the Alpine pastures as one herd; the cheese produced from the sweet-tasting milk was distributed among the owners. The Togg was imported into the British Isles in 1822 and the following year four were exported to the USA. This robust breed has been exported to many countries, but like most Alpine breeds is happier in temperate climes.

Their medium-length coat is light fawn to the darkest chocolate; the ears are white with a dark spot in the middle and two white stripes down the face. The lower parts of the legs are white. In the *Guinness Book of World Records* the Toggenburg owned by Lilian Sandburg of North Carolina holds the record for producing 1,140 gallons (5,182.5 litres) of milk in 365 days.

36

Characteristics

Weight: male 132–198 lb (60–90 kg), female 99–132 lb (45–60 kg). Height: male 29½ in (75 cm), female 29½ in (75 cm). Scimitar-shaped horns: billy 31½ in (80 cm) long, nanny 17¾ in (45 cm) long.

Valais Blackneck

The Valais Blackneck is a dual-purpose breed with an international reputation and is known by many names, including Col Noir de Valais and the Glacier goat. The Valais Blackneck was developed from indigenous goats crossed with the Italian Kupferziege goat and further improved by selective breeding, and are found mainly in the Lower Valais in Switzerland. This hardy breed has long, shaggy, wavy hair, white from the shoulders back; the head, neck and front legs are black. They have the ability to tolerate the harsh winter conditions but are housed in the worst weather. It is a gourmet browser feeding only on the greenest of Alpine grass, fresh herbs and flowers and producing up to 4¼ pints (2 litres) of milk daily.

In 1970 this protected species was close to extinction, but fortunately for this most beautiful, most photographed goat the numbers are improving, and there are now approximately three thousand.

Acknowledgements

I would like to acknowledge the help and advice received from the following people that was crucial in putting together the first edition of this book: Ólafur R. Dýrmundsson PhD, National Adviser on Organic Farming and Land Use, Iceland; Tony Harman of Maple Leaf Images, Skipton, North Yorkshire for photographic help and advice; Elaine my wife, my daughters Ruth and Karen and my granddaughter Rebecca for all their help.

Picture Credits

Plate (1) Margot Wolfs; (2) Ian Preston; (3) Christoph Ulrich; (4) Michael Trotter; (5) The Bagot Goat Society; (6) Bilberry Goat Trust; (7) Mariagrazia Arrighini; (8) Zulkifli Ishak; (9) Paul Mounter; (10) Clive Dodd; (11) www.plamp.cz; (12) Martin Kaufmann; (13) Wibe-Jan Postma; (14) Candi Morgan McCorkle, Rustling Oaks Farm; (15) Salvatore Pipia; (16) Steve Bates; (17) Johanna B. Thorvaldsdottir, Haafell Geitfjarsetur; (18) Hancock Kiko Farm; (19) Kelsee Gibbs at Kinder Korner Goats; (20) Lance Hays; (21) Universita di Palermo; (22) Courtesy of ANGBA © Bessie Miller; (23) Erin Hottle; (24) Gillian Cunningham, Willow Tree Farm; (25) Beate Milerski; (26) Levend Landgoed NOVA; (27) Hilary Breakell, Marshview Viggo; (28) Little Bit Acres; (29) Franck Rimaud; (30) Brian Goodwin; (31) Jeannette Beranger; (32) © Lindy Warner Photography; (33) Morgan Fredericks; (34) Michèle Hennin; (35) Willowbank Toggenburgs; (36) Paul Asman and Jill Lenoble.

Shutterstock: Volkova Irina (cover); Pierluigi.Palazzi (pp. 2-3); Artsiom Petrushenka (p. 6)